This book belongs to:

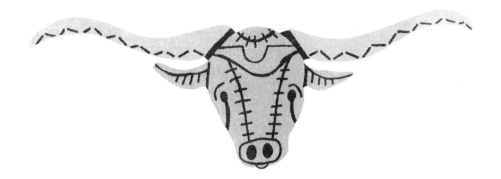

HOW
TO BE A
COWBOY

Alice V. Lickens

PAVILION

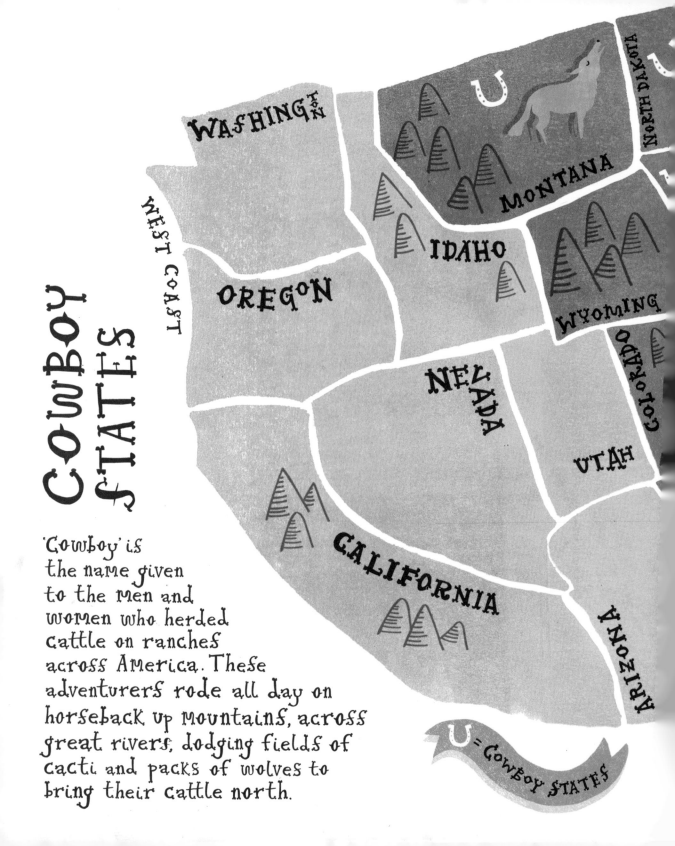

COWBOY STATES

WEST COAST

WASHINGTON

OREGON

IDAHO

MONTANA

NORTH DAKOTA

WYOMING

COLORADO

NEVADA

UTAH

CALIFORNIA

ARIZONA

'Cowboy' is the name given to the men and women who herded cattle on ranches across America. These adventurers rode all day on horseback up mountains, across great rivers, dodging fields of cacti and packs of wolves to bring their cattle north.

U = COWBOY STATES

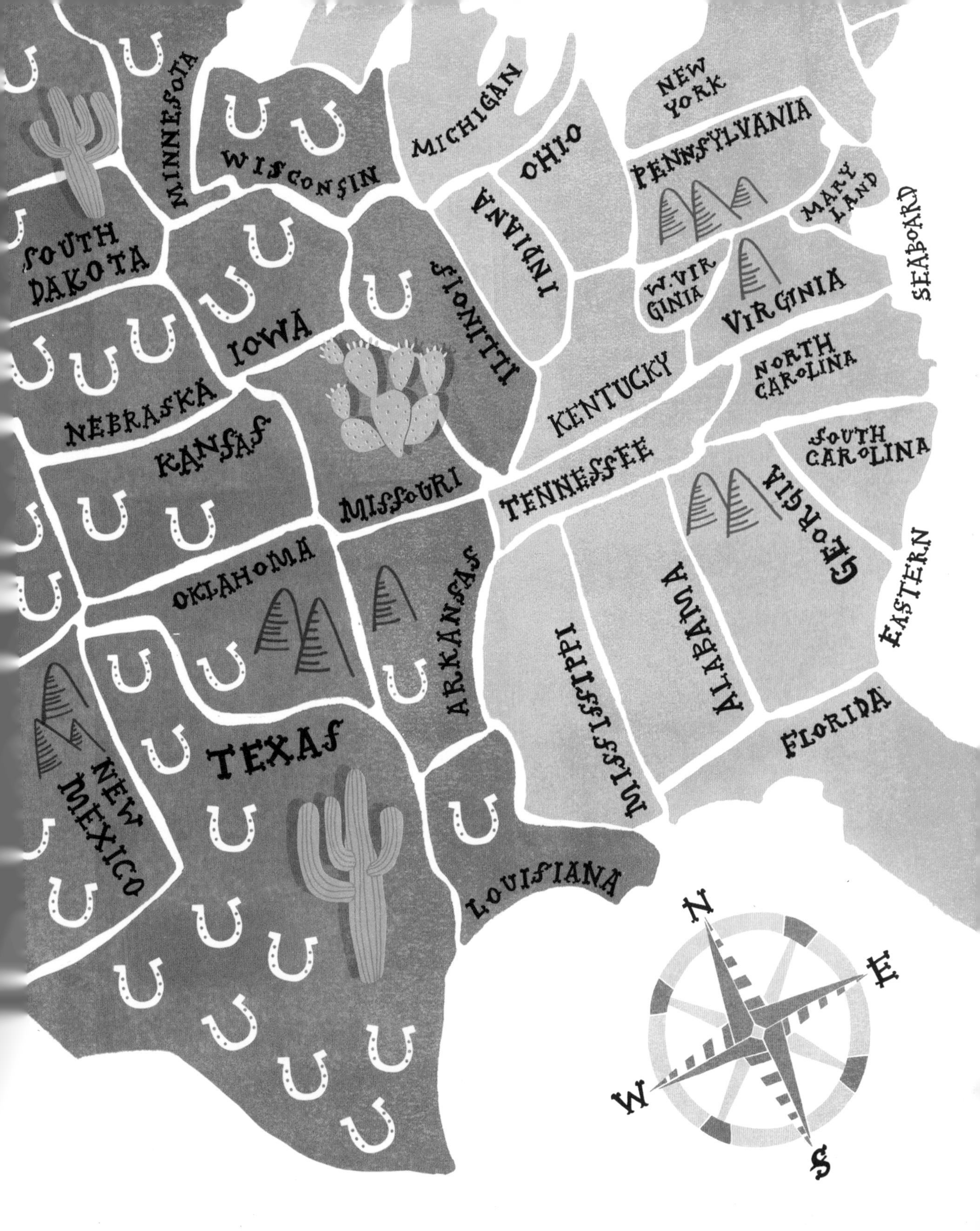

MINNESOTA
WISCONSIN
MICHIGAN
NEW YORK
PENNSYLVANIA
OHIO
INDIANA
MARY LAND
SEABOARD
SOUTH DAKOTA
IOWA
ILLINOIS
W. VIR GINIA
VIRGINIA
NEBRASKA
KANSAS
KENTUCKY
NORTH CAROLINA
SOUTH CAROLINA
MISSOURI
TENNESSEE
GEORGIA
OKLAHOMA
ARKANSAS
ALABAMA
MISSISSIPPI
EASTERN
FLORIDA
TEXAS
NEW MEXICO
LOUISIANA
N
E
W
S

BARN

SHED

On the Ranch

The ranch is the cowboy's base and home and where they head off from on their long trail drives.

COOKHOUSE

BUNKHOUSE

RANCH
HOUSE

STABLES

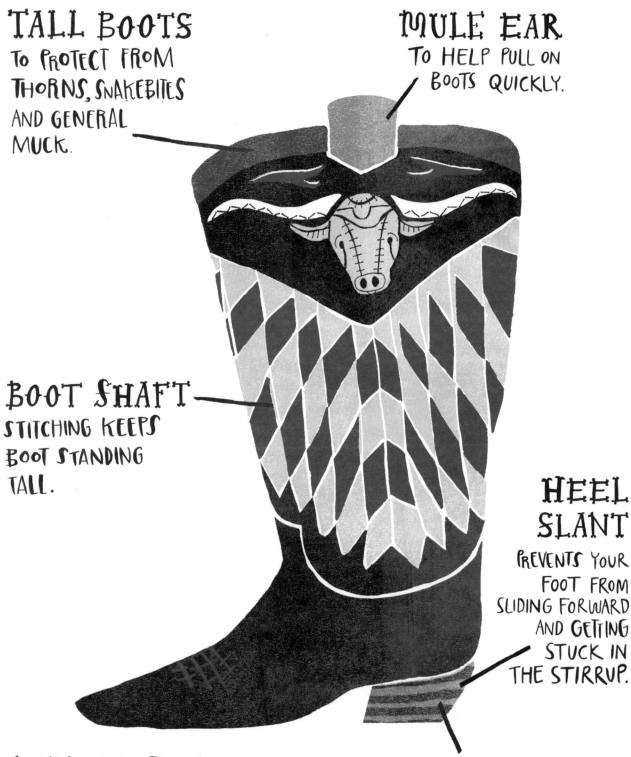

TALL BOOTS
TO PROTECT FROM THORNS, SNAKEBITES AND GENERAL MUCK.

MULE EAR
TO HELP PULL ON BOOTS QUICKLY.

BOOT SHAFT
STITCHING KEEPS BOOT STANDING TALL.

HEEL SLANT
PREVENTS YOUR FOOT FROM SLIDING FORWARD AND GETTING STUCK IN THE STIRRUP.

NARROW HEEL TO DIG INTO THE DIRT WHEN LASSOING A STEER.

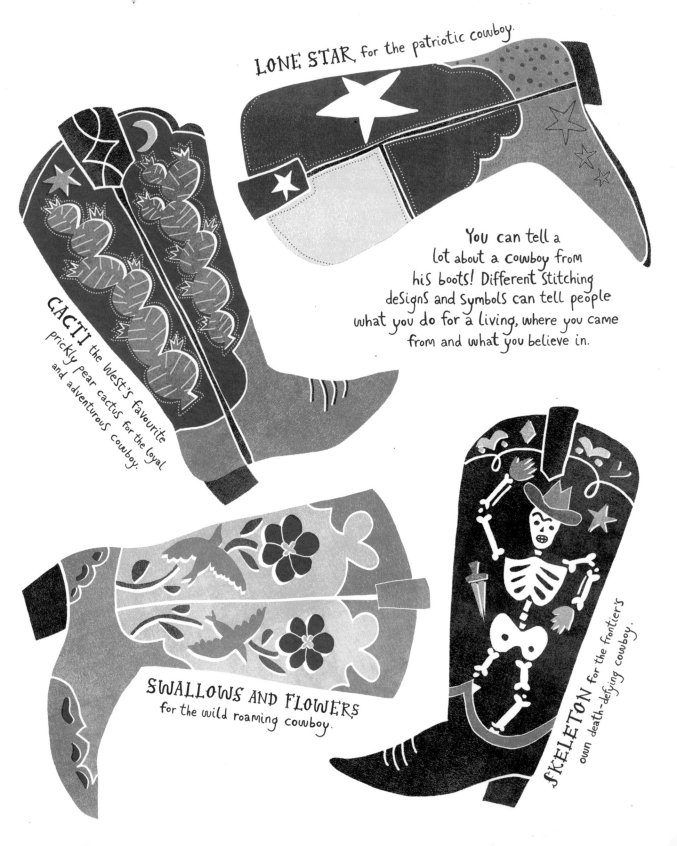

LONE STAR for the patriotic cowboy.

You can tell a lot about a cowboy from his boots! Different stitching designs and symbols can tell people what you do for a living, where you came from and what you believe in.

CACTI the West's favourite prickly pear cactus for the loyal and adventurous cowboy.

SWALLOWS AND FLOWERS for the wild roaming cowboy.

SKELETON for the frontier's own death-defying cowboy.

A COWBOY NEEDS A RANGE NAME TO BE SHOUTED ACROSS THE PRAIRIE. WHAT WOULD YOURS BE?

SKEETER WHAM

CODY CLUSTER

BLUE CLOUD WRIGHT

SILVER WOLFSCALE

ROWDY BILL

VALENTINO CURLEY

RED BENSON

ROCKY HORNCRACKLE

JESSE BIRCH

GUMBO DOBSON

COTTON COKENDALE

PEE WEE LITTLE

SCOUT ZOOP

LU TOUNT

PENNY LEWIS

BOBCAT SWEETWATER

POW CARTER

JJ BUCK

BOO MIX

ACE BERRY

BOON MUDD

BUBBA HUMBLE

HOOTER CLARENCE

RUSTY MOUNT

ZEKE FLOSS

MADS CHISUM

KID CASSIDY

DARLA DOOLIN

STETSONS

'cause you can't be a Cowboy without a hat!

10-GALLON

TEXAS

SUGARLOAF SOMBRERO

Your hat keeps you cool, defends you against cacti, is a water bucket for you and your steed, hides your poker winnings, keeps you snug on a snowy night and makes you a king on the plains!

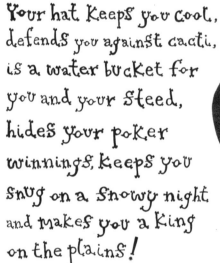

PLAINSMAN

BOSS OF THE PLAINS

Cowboy hats are over 150 years old and are made of wool felt and beaver fur.

MONTANA PEAK

SHERIFF

MEET YOUR MUSTANG!

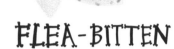

PINTO

Large patches of white with splotches of black and chestnut.

FLEA-BITTEN

Grey or roan with specks of black and chestnut.

Fuzztail
a wild
mustang horse

APPALOOSA

Leopard-spotted coat, bred from Native American horses.

Broomtail
a wild
mustang mare

Bronco
a wild
horse

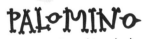

GR...
Mouse-col...

PALOMINO

Gold coat with white mane and tail.

PITCHFORK

RUNNING M

BOX T

ROCKING H

FLYING U

RUNNING F

PIG PEN

BARBECUE

O-IN-A-HOLE

T-DOWN BAR

SEVEN UP

CIRCLE DIAMOND

TWO-POLE PUMPKIN

SPADE

ROCKING A

LAZY 9

Y CROSS

DOUBLE H

TRIPLE V

WALKING X

TUMBLING Y

LAZY M

RAFTER O

M BAR J

A REAL COWBOY CAN READ HIS CATTLE LIKE A BILLBOARD.

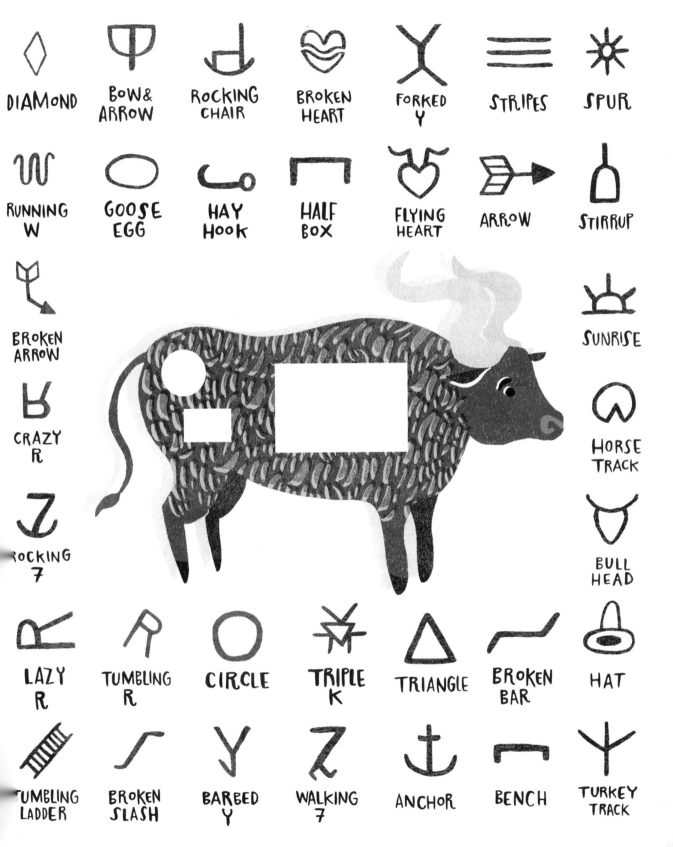

DIAMOND

BOW & ARROW

ROCKING CHAIR

BROKEN HEART

FORKED Y

STRIPES

SPUR

RUNNING W

GOOSE EGG

HAY HOOK

HALF BOX

FLYING HEART

ARROW

STIRRUP

BROKEN ARROW

CRAZY R

ROCKING 7

SUNRISE

HORSE TRACK

BULL HEAD

LAZY R

TUMBLING R

CIRCLE

TRIPLE K

TRIANGLE

BROKEN BAR

HAT

TUMBLING LADDER

BROKEN SLASH

BARBED Y

WALKING 7

ANCHOR

BENCH

TURKEY TRACK

OLD BLUE MT.

BLUE MT.

CAPOTE MT.

SAWTOOTH MT.

STAR MT.

ELEPHANT MT.

Texas Longhorns are the best cattle to take out on the range as they are uniquely adapted to survive the long, treacherous journey north. They can run like deer, swim like fish and fight off the meanest wolf.

Descended from the cattle brought to the USA by Christopher Columbus, these beasts are lean and tough with enormous horns.

LOST MINE PEAK

EL CAPITAN

CATHEDRAL MT.

Old Blue was a trusty longhorn so experienced that, for many years, he led the herd along the trail. He wore a bell round his neck so the cattle could follow him.

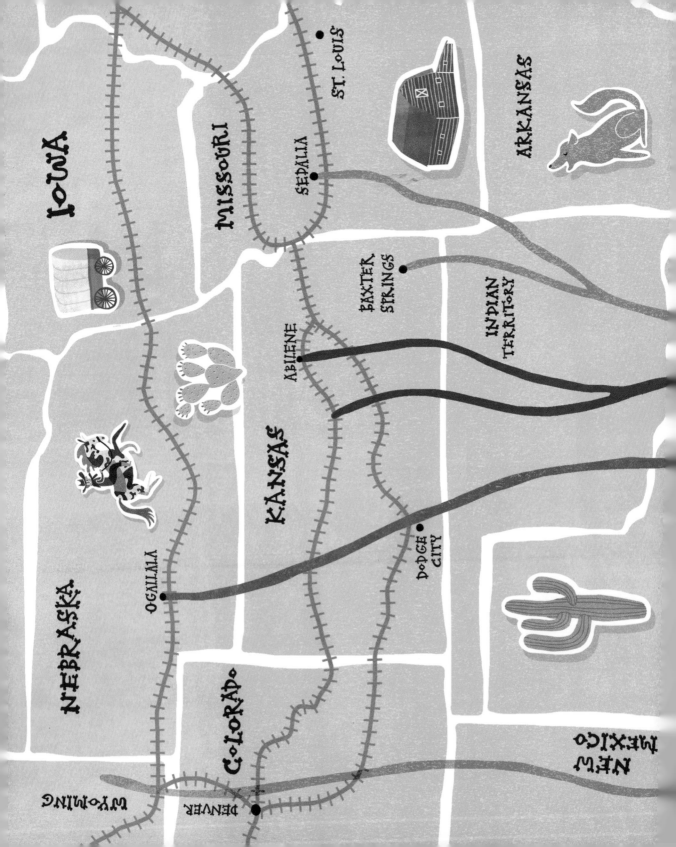

LOUISIANA

TEXAS

FORT CONCHO

CATTLE TRAILS

A group of ten cowboys could herd up to 3,000 cattle up through the rough terrain of the plains on a 1,000 mile journey north. Cowboys would travel the fairly short distance of 15 miles a day and give the herd regular breaks to graze so the cows get too skinny on the journey. It could tal up to two months to travel these trails!

REMUDA

The extra horses travelling with the cowboys.

WRANGLER

The youngest and greenest cowboy is in charge of the remuda.

DRAG

The worst job on the trail: these riders keep the cattle moving forward and ride in the dust cloud kicked up by the herd.

COWBOY FORMATION

To keep the herd moving smoothly each cowboy has a specific job and position on the trail.

FLANK

SWING

CHUCK WAGON would be driven by the cookie and would push ahead of the pack to make camp for the cowboys at the end of the day's drive.

LEADER A reliable cow who marches at the head of the herd.

TRAIL BOSS Rides in front, scouting the way.

POINT Rides at the front of the cattle and directs the herd movement.

wing and Flank riders travel at the Side of the herd preventing the cattle from Spreading out.

TALK LIKE A COWBOY!

A hog killin'-time ~ a really good time

Beanslinger ~ the cook

Calaboose ~ jail

Big Bug ~ the ranch boss

By gum! ~ an inoffensive cuss

Jackaroo ~ a cowboy

Old Dan ~ a trusty old horse

Two whoops and a holler ~ not far away

Trollybobs ~ entrails

Pernsnickity ~ to be unusually picky

West Texas rain ~ dust

GUNSLINGER BEANS

INGREDIENTS

2 lb Pinto Beans
2 lb Ham Hock
(or Salt Pork)
2 Onions, chopped

4 Tablespoons Sugar
2 Green Chillies
Can Tomato Paste

Wash beans and soak overnight. Drain, place in a Dutch oven and cover with water. Add remaining ingredients and simmer until tender. Sample the beans while cooking. Add salt to taste.

Home on the Range

A chuck wagon was a modified cart invented by famous rancher Charles Goodnight in 1886. It would carry all the provisions and tools needed by the cowboys. Most importantly, the chuck wagon had a specially designed set of drawers and shelves to store all the cooking essentials.

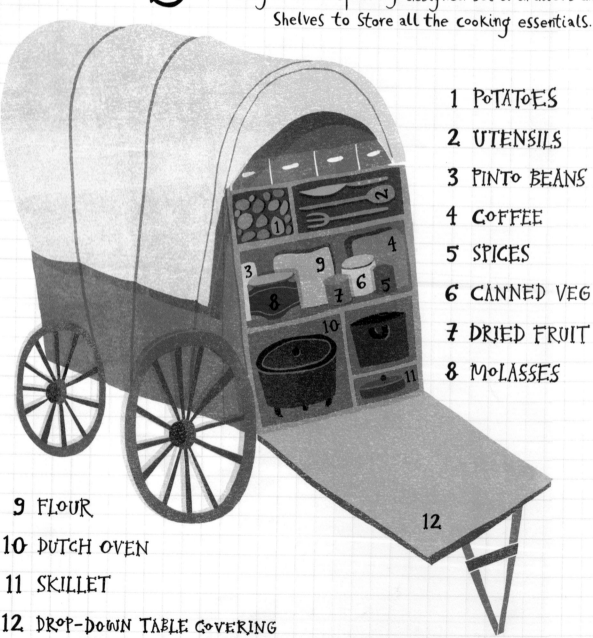

1 POTATOES

2 UTENSILS

3 PINTO BEANS

4 COFFEE

5 SPICES

6 CANNED VEG

7 DRIED FRUIT

8 MOLASSES

9 FLOUR

10 DUTCH OVEN

11 SKILLET

12 DROP-DOWN TABLE COVERING THE CHUCK SHELVES

Grub pile ~ a meal from the chuck wagon

Rattle your hocks ~ hurry up

Dutch oven ~ a sturdy three-legged cooking pot

Chucklehead ~ a fool

Owl hoot trail ~ an outlaw's way of life

Chickabiddy ~ a young chicken or child

Spudgel ~ to run away

Vamoose ~ to get going

Flumguzzling ~ to deceive someone

Zitted ~ moving so fast you're flying

Mad as a hornet ~ very mad

Puddin' foot ~ an awkward horse

SON OF A GUN STEW

INGREDIENTS

2 lb Lean Beef
Half a Calf Heart
1 ½ lb Calf Liver
1 Set Sweetbreads

1 Set Brains
1 Set Marrow Gut
Louisiana Hot Sauce
Salt and Pepper

Slice the marrow gut into small rings. Add salt and pepper to taste followed with hot sauce. Cover meat with water and simmer for 2-3 hours. Take sweetbreads, heart, brains, and liver and cut into small pieces. Add to stew. Simmer and boil. Place in a Dutch oven or deep casserole.

Before bed the Cookie uses the tongue of his wagon to point to the North Star. In the morning the Trail Boss would use this makeshift compass to guide the herd onwards.

Cowboys like their coffee strong, 'so thick it would float a horseshoe'.

Look like a cowboy - open the flaps!

Cowboys carried canned tomatoes if they had a long solitary ride as they were better at staving off thirst than water.

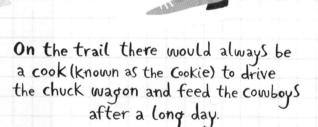

On the trail there would always be a cook (known as the Cookie) to drive the chuck wagon and feed the cowboys after a long day.

A Cookie's most prized possession is his sourdough starter (a living yeast paste) used to make bread, biscuits and other tasty snacks. To keep your starter alive you must keep it warm so at night the Cookie tucks it into his bedroll with him.

ROUNDUP STEW

INGREDIENTS

6 Potatoes
6 Carrots
2 lb Beef

1 Onion
Salt and Pepper

Chop meat and brown in a Dutch oven. Cover with water and simmer for one hour or more. Add salt and pepper, then the veg and cook for a further 30 minutes.

Serve!

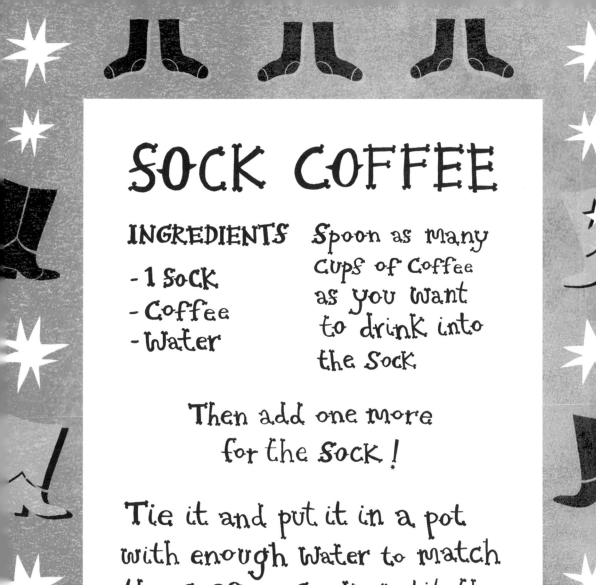

SOCK COFFEE

INGREDIENTS

- 1 Sock
- Coffee
- Water

Spoon as many cups of Coffee as you want to drink into the Sock

Then add one more for the Sock!

Tie it and put it in a pot with enough water to match the Coffee. Cook until the Sock is the Colour of your dirtiest sock.

Serve!

These smart and gentle cattle are driven on the trail. Their horns can grow up to 2 metres long.

LONGHORN

COYOTE

Small wild canines. The name of these cunning animals derives from the Aztec word for 'trickster.'

ARMADILLO

Small critter with a leathery armoured shell, occasionally good for a barbecue.

Jackalope

This (purely fictitious) terrifying critter is a favourite from plains' folklore – half-jackrabbit, half-antelope.

BUFFALO

The heaviest animal in North America, they can run up to 40 mph.

ROADRUNNER

Small and quick on their feet, these birds eat lizards, insects and the occasional rattlesnake.

PRAIRIE CHICKEN

These large roaming grouse once thrived in the plains, but are now extremely rare.

PEGASUS

ORION

CANCER

TO GUIDE THEM HOME COWBOYS RODE BY THE STARS

For my
brother
Joe

THIS EDITION FIRST PUBLISHED IN THE
UNITED KINGDOM IN 2015 BY
PAVILION CHILDREN'S BOOKS,
AN IMPRINT OF PAVILION BOOKS GROUP LIMITED,
1 GOWER STREET
LONDON
WC1E 6HD

LAYOUT © PAVILION BOOKS, 2015
ILLUSTRATIONS AND TEXT © ALICE V LICKENS, 2015

THE MORAL RIGHTS OF THE AUTHOR AND ILLUSTRATOR HAVE BEEN ASSERTED.

DESIGNER: CLAIRE CLEWLEY
PRODUCTION CONTROLLER: LAURA BRODIE
COMMISSIONING EDITOR: KATIE DEANE

ISBN: 9781843652410

A CIP CATALOGUE RECORD FOR THIS BOOK IS AVAILABLE FROM THE BRITISH LIBRARY.

10 9 8 7 6 5 4 3 2 1

REPRODUCTION BY MISSION PRODUCTIONS, HONG KONG
PRINTED AND BOUND BY 1010 PRINTING INTERNATIONAL LTD, CHINA

THIS BOOK CAN BE ORDERED DIRECTLY FROM THE PUBLISHER ONLINE AT WWW.PAVILIONBOOKS.COM